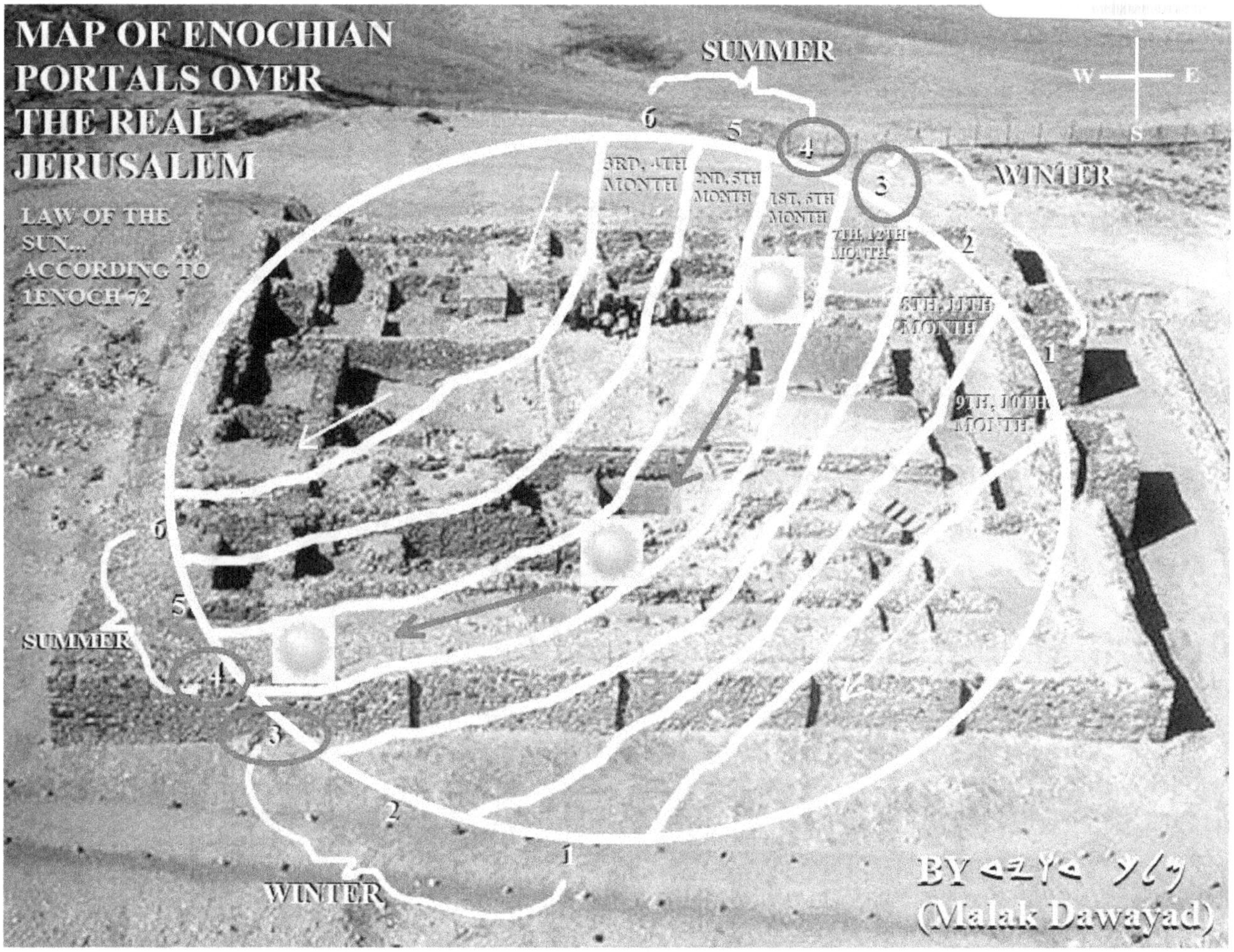

THE UNIVERSAL ENOCH CALENDAR FROM THE REAL JERUSALEM IN TEL ARAD, ISRAEL
BY ◁ΖΥ◁ Ƴ⌐Ƴ (Malak Dawayad)

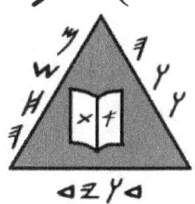

Table of Contents

Ten Important Rules – pg.3-6

Monthly Structure – pg.7-10

Lifetime Calendar – pg.10

Sabbath Year Calendars – pg.11

Enochian Zodiac & Jubilee Calendars – pg.12-15

Actual Enoch Calendar – pg.16-27

Supportive Sources – pg.28

Donate – pg.29

Contact – pg.30

10 IMPORTANT RULES

1.) 12-MONTH CALENDAR NOT 13 MONTHS...

1Ki 4:7 And Solomon had twelve officers over all Israel, which provided victuals for the king and his household: each man his month in a year made provision.

1Chr 27:1-15 Now the children of Israel after their number, *to wit,* the chief fathers and captains of thousands and hundreds, and their officers that served the king (David) in any matter of the courses, which came in and went out month by month throughout all the months of the year, of every course *were* twenty and four thousand. (2) Over the first course for the first month *was* Jashobeam the son of Zabdiel: and in his course *were* twenty and four thousand... (15) The twelfth *captain* for the twelfth month *was* Heldai the Netophathite, of Othniel: and in his course *were* twenty and four thousand.

Enoch 82:11 Their four leaders who divide the four parts of the year (seasons) enter first; and after them the twelve leaders of the orders who divide the months (12); and for the three hundred and sixty (days) there are heads over thousands who divide the days; and for the four intercalary days there are the leaders which sunder the four parts of the year.

Rev 22:2 ...*was there* the tree of life, which bare twelve *manner of* fruits, *and* yielded her fruit every month: and the leaves of the tree *were* for the healing of the nations.

2.) THE NEW YEAR IS EXCLUSIVELY THE 1ST DAY OF WEEK

JASHER 83:1-4
And on the eighth day, being the first day of the first month, in the second year from the Israelites' departure from Egypt... *(Remembrance of Enoch 4:51) (2Enoch 33:1)*

JUBILEES 7:2
And he made wine therefrom and put it into a vessel, and kept it until the fifth year, until the first day, on the new moon of the first month.

Legends of the Jews
Book III, Chapter 3:Day of Ten Crowns ...it was besides the first day of the week, the first day of the first month of the year.

IMPORTANT RULES

*3.) **THE 8TH DAY IS EXCLUSIVELY THE 1ST DAY OF THE WEEK***

Remembrance of Enoch 4:51 …the eighth day or the first day of the week…

2Enoch 33:1 And I appointed the eighth day also, that the eighth day should be the first- created after my work…

*4.) **364 DAYS COMPLETES A FULL YEAR***

JUBILEES 6:32
And command thou the children of Israel that they observe the years according to this reckoning- three hundred and sixty-four days, and (these) will constitute a complete year, and they will not disturb its time from its days and from its feasts; for everything will fall out in them according to their testimony, and they will not leave out any day nor disturb any feasts.

ENOCH 82:6-7

… and the year is completed in three hundred and sixty-four days. And the account thereof is accurate and the recorded reckoning thereof exact…

*5.) **The 1st day of the 2nd, 5th, 8th & 11th months will always be the 3rd day of the week & the 1st day of the 3rd, 6th, 9th & 12th months will always be the 5th day of week***

Book of the Bee 20: last verse …He died on the fourth day of the week, on the second of Nîsân? (Iyar/2nd month), at the second hour of the day; his son Shem embalmed him, and his sons buried him, and mourned over him forty days.

Chronicles of Jerahmeel 61:1 The banishment brought about by Titus, Vespasianus, and Hadrian, occurred on the eve of the ninth of Ab (6th month), on the outgoing of the Sabbath (6th day of the week) and the Sabbatical year…

IMPORTANT RULES

6.) MONTHS HAVE 30 DAYS (360 TOTAL) + 1 DAY TO COMPLETE EACH SEASON (4)

1Enoch 82:4, 15 Blessed are all the righteous, blessed are all those who walk In the way of righteousness and sin not as the sinners, in the reckoning of all their days in which the sun traverses the heaven, **entering into and departing from the portals for thirty days** with the heads of thousands of the order of the stars, together with **the four which are intercalated which divide the four portions of the year**, which lead them and **enter with them four days.** In the beginning of the year, Melkiel rises first and rules, who is called the southern Sun - and all the days of his period, during which he rules, are ninety-one.

Jubilees 6:28-29 And on this account he ordained them for himself as feasts for a memorial for ever, and thus are they ordained. And they placed them on the heavenly tables, each (season) had **thirteen weeks (13 x 7=91 days)**; from one to another (passed) their memorial, from the **first to the second, and from the second to the third, and from the third to the fourth (91 DAYS X 4 SEASONS= 364 DAYS).**

7.) Days of the Week, Month & Year are NOT Determined by Moon Phases

Jubilee 6:35-38 ...lest they forget the feasts of the covenant and walk according to the feasts of the Gentiles after their error and after their ignorance. For there will be those who will assuredly **make observations of the moon--now (it) disturbeth the seasons and cometh in from year to year ten days too soon.** For this reason **the years will come upon them when they will disturb (the order)**, and make an abominable (day) the day of testimony, and an unclean day a feast day, and they will confound all the days, the holy with the unclean, and the unclean day with the holy; for they will go **wrong as to the months and sabbaths** and feasts and jubilees.

2Enoch 16:3 ...while the lunar year has **three hundred fifty-four days**

8.) Daily Succession of all 12 months are 30 days, 30 days & then 31 days for every 3 months, starting with the 1st, 2nd & 3rd month, then continue the same pattern

1Enoch 72:6-14 In this way he rises in **the first month** in the great portal, which is the fourth [those six portals in the east]. And in that fourth portal from which the sun rises in the first month are twelve window-openings, from which proceed a flame when they are opened in their season. When the sun rises in the heaven, he comes forth through that fourth portal **thirty, mornings in succession**, and sets accurately in the fourth portal in the west of the heaven. And during this period the day becomes daily longer and the night nightly shorter to the thirtieth morning. On that day the day is longer than the night by a ninth part, and the day amounts exactly to ten parts and the night to eight parts. And the sun rises from that fourth portal, and sets in the fourth and returns to the fifth portal of the east **thirty mornings (2nd month),** and rises from it and sets in the fifth portal. And then the day becomes longer by two parts and amounts to eleven parts, and the night becomes shorter and amounts to seven parts. And it returns to the east and enters into the sixth portal, and rises and sets in the sixth portal **one-and-thirty mornings (3rd month),** on account of its sign. On that day the day becomes longer than the night, and the day becomes double the night, and the day becomes twelve parts, and the night is shortened and becomes six parts.

IMPORTANT RULES

*9.) **All Dates Begin @Twilight, NOT at 12am/Midnight or Sunrise***

Job 3:9 Let the stars of the twilight thereof be dark; let it look for light, but *have* none; neither let it see the dawning of the day:

Neh 4:21 So we laboured in the work: and half of them held the spears from the rising of the morning till the stars appeared.

*10.) **Days, Weeks, Months & Years are Calculated by the Sun***

Jubilees 2:9 And God appointed the sun to be a great sign on the earth for days and for sabbaths and for months and for feasts and for years and for sabbaths of years and for jubilees and for all seasons of the years.

Chronicles of Jerahmeel 4:9 The sun is appointed over the light, to separate light from darkness, and through it to enable us to calculate the days, months and years, and to do every kind of work, to walk any distance, and to migrate from city to city and from town to town...

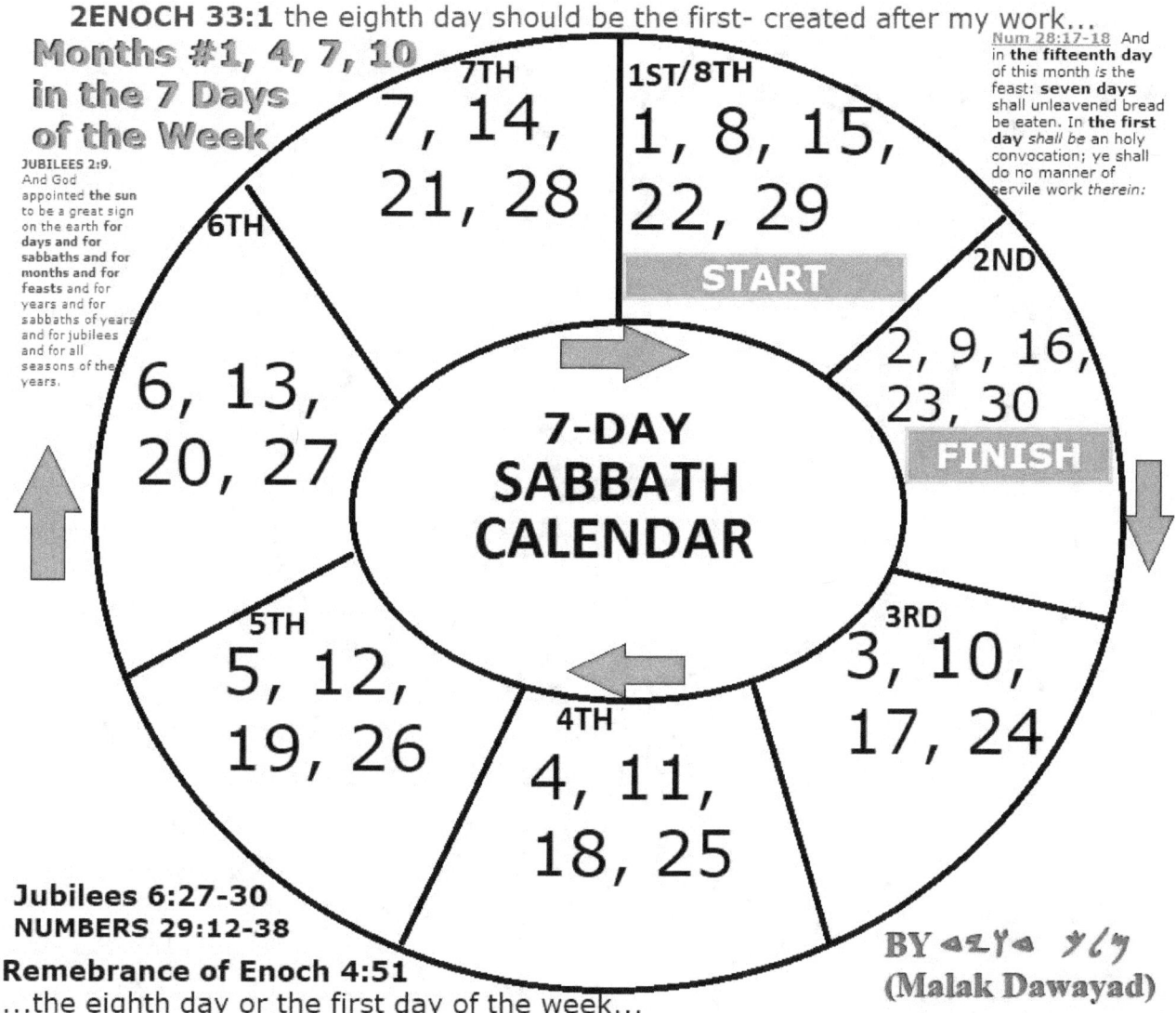

Structure of Months #1, 4, 7 and 10

The 1st, 4th, 7th, & 10th month always begin on the 1st day of the week

Num 28:16-18 And in the fourteenth day (Sabbath day) of the first month is the passover of the LORD. And in the fifteenth day of this month is the feast: seven days shall unleavened bread be eaten. In the first day shall be an holy convocation; ye shall do no manner of servile work therein:

The 14th day of the 1st month which is divisible by 7 is obviously a Sabbath day. The 15th day being the beginning of the feast of Unleavened Bread falling on the 1st day of the week also proves the 1st day of the month will always be on the 1st day of the week. We've already established that the first month, the fourth month, seventh month and as well as the tenth month all follow the four intercalary days which fall on the Sabbath day. So, it's already understood that these months would start on the first day of the week.

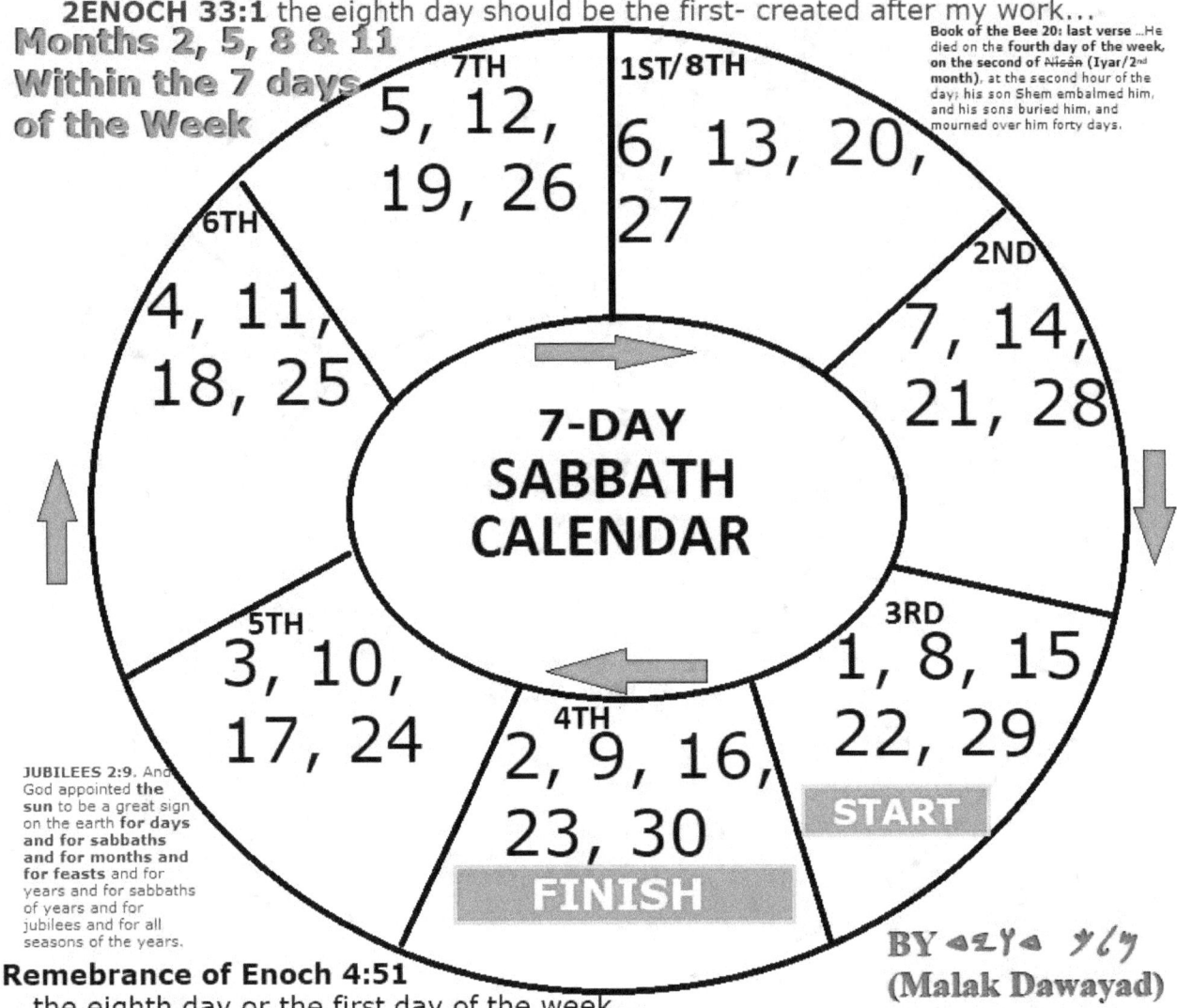

Structure of Months #2, 5, 8 and 11
The 2nd, 5th, 8th, & 11th month always begin on the 3rd day of the week.

Remember the monthly succession of days are 30 days, 30 days & then 31 days for every 3 months, starting with the 1st, 2nd & 3rd month according to 1Enoch 72:6-35. The 1st month starting on the 1st day of the week has 30 days, so the last Sabbath of the month would be the 28th day (divisible by 7) and 3 days later on the third day of the week would be the 1st day of the 2nd month. The 2nd month and likewise the 5th, 8th, & 11th month would all start on the 3rd day of the week.

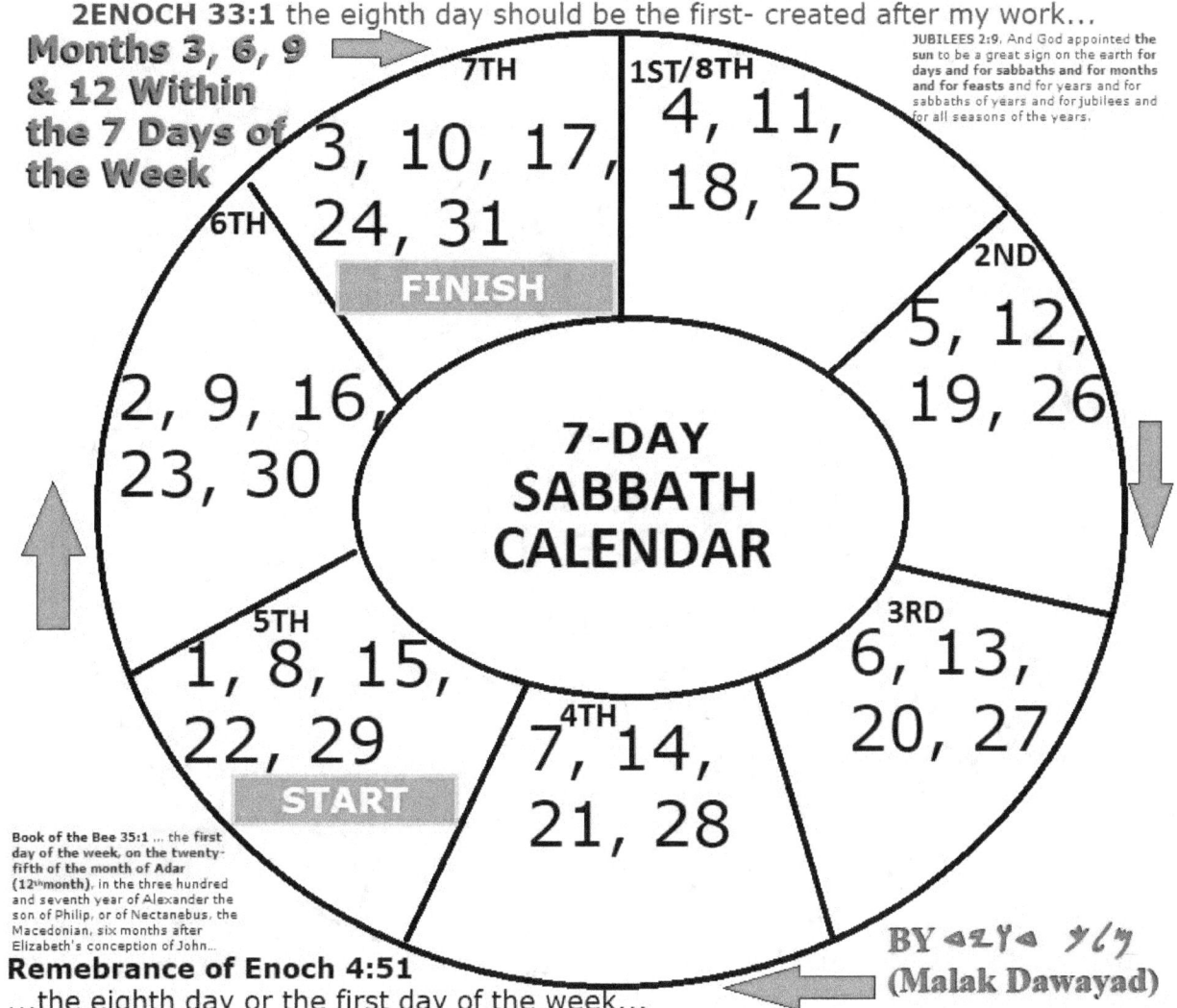

Structure of Months #3, 6, 9 and 12
The 3rd, 6th, 9th, & 12th month always begin on the 5th day of the week.

Book of the Bee 35:1 At the ninth hour of the first day of the week, on the twenty-fifth of the month of Adar (12th month), in the three hundred and seventh year of Alexander the son of Philip, or of Nectanebus, the Macedonian, six months after Elizabeth's conception of John...

The sequence of Sabbath days in the 3rd, 6th, 9th & 12th month is the 3rd day, 10th, 17th, 24th & 31st day of the month. Knowing that, the Book of the Bee 35 says the 25th day of the 12th month falls on the 1st day of the week. This proves the 12th month begins on the 5th day of the week.

2ENOCH 33:1 the eighth day should be the first- created after my work...

○ 1st, 4th, 7th & 10th month
○ 2nd, 5th, 8th & 11th month
● 3rd, 6th, 9th & 12th month

Psa 90:12 So teach *us* to **number our days,** that we may apply *our* hearts unto wisdom.

7-DAY SABBATH CALENDAR
(Lifetime Calendar)

1ST/8TH
1, 8, 15, 22, 29
6, 13, 20, 27
4, 11, 18, 25
Start of the 1st, 4th, 7th & 10th Month
START

2ND
2, 9, 16, 23, 30
7, 14, 21, 28
5, 12, 19, 26

3RD
3, 10, 17, 24
1, 8, 15, 22, 29
6, 13, 20, 27
Start of the 2nd, 5th, 8th & 11th Month
START

4TH
4, 11, 18, 25
2, 9, 16, 23, 30
7, 14, 21, 28

5TH
5, 12, 19, 26
3, 10, 17, 24
1, 8, 15, 22, 29
Start of the 3rd, 6th, 9th & 12th month
START

6TH
6, 13, 20, 27
4, 11, 18, 25
2, 9, 16, 23, 30

7TH
7, 14, 21, 28
5, 12, 19, 26
3, 10, 17, 24, 31

Remebrance of Enoch 4:51
...the eighth day or the first day of the week...

BY ⟨⟨⟨⟩⟩⟩
(Malak Dawayad)

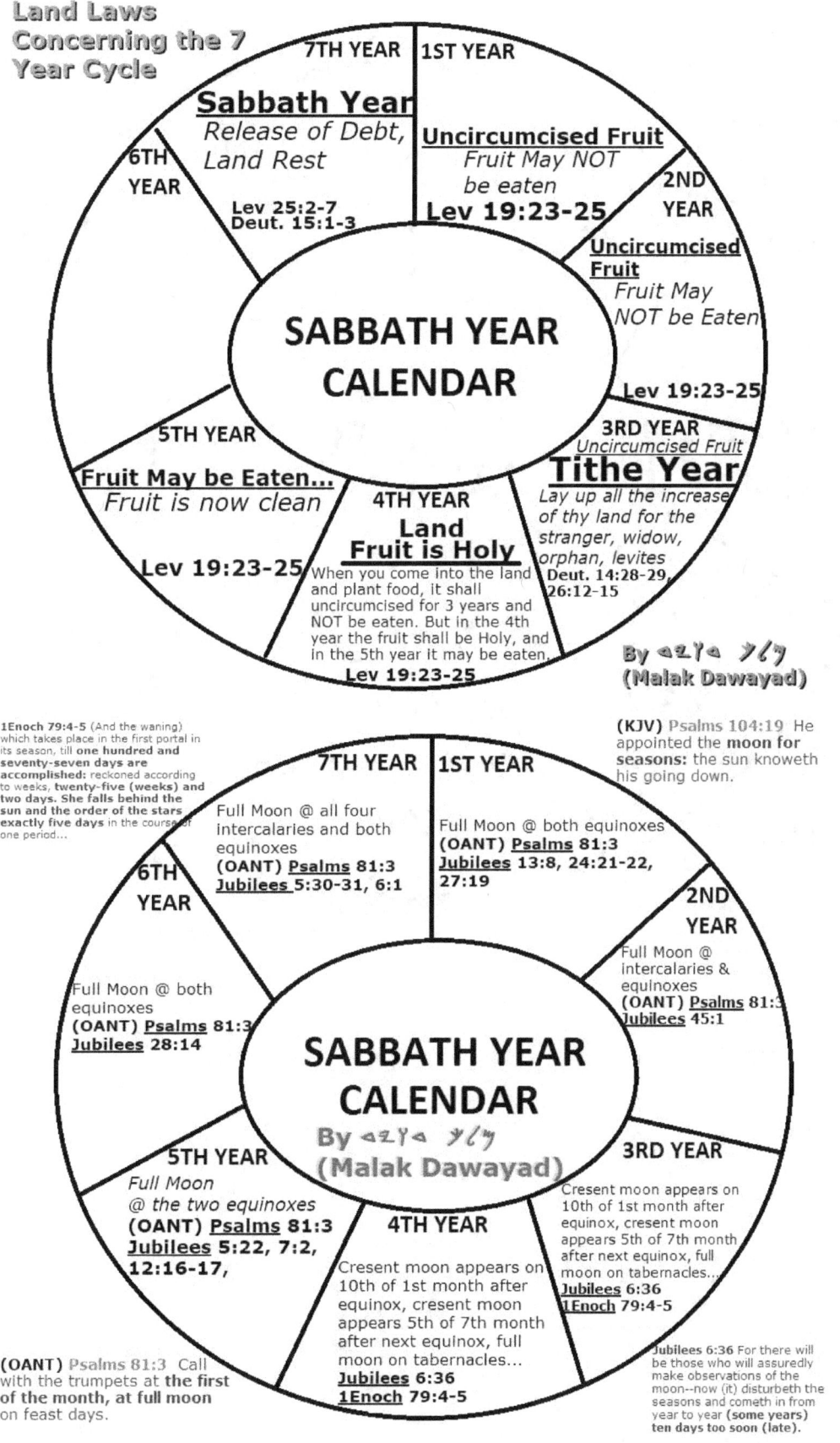

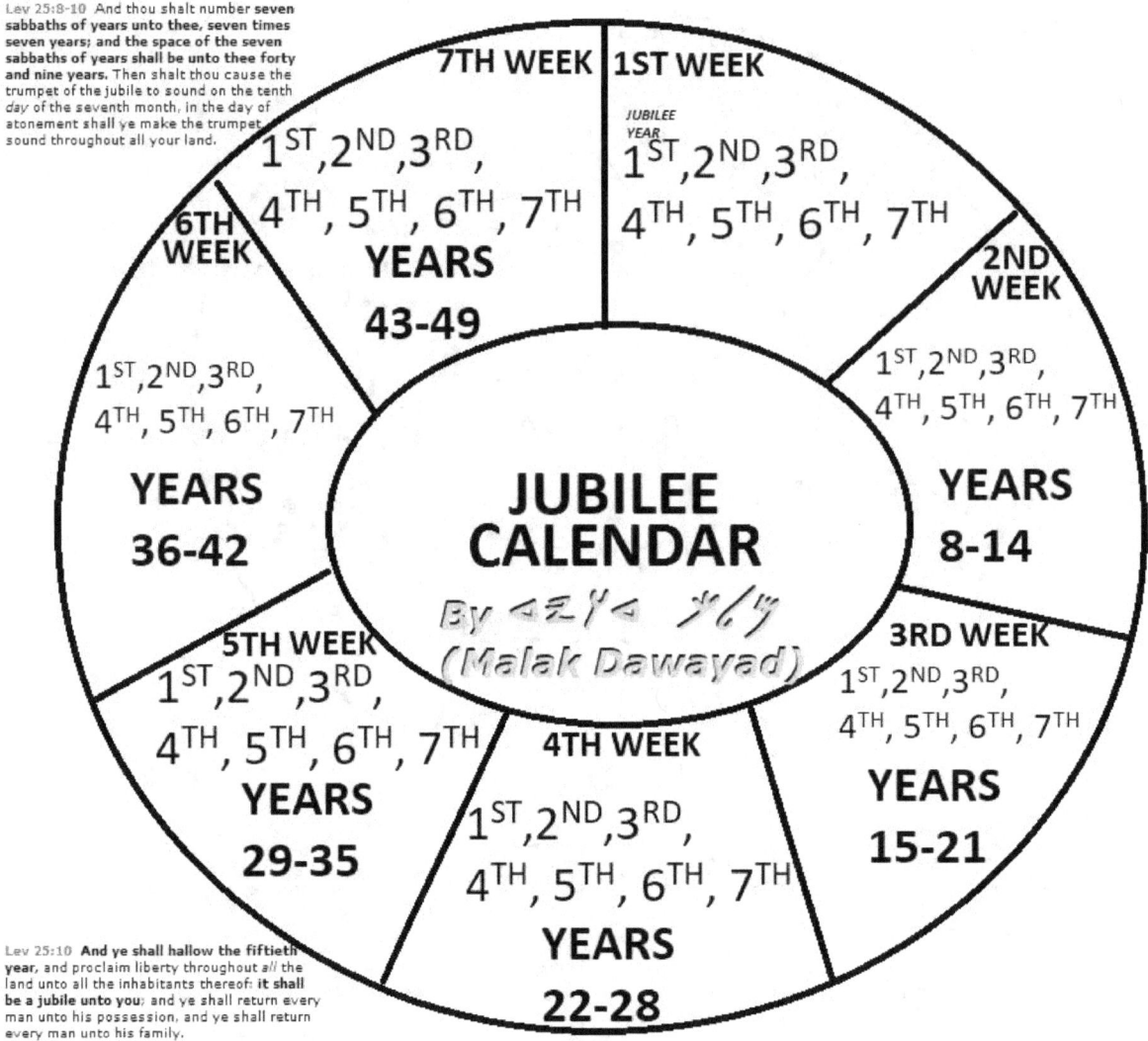

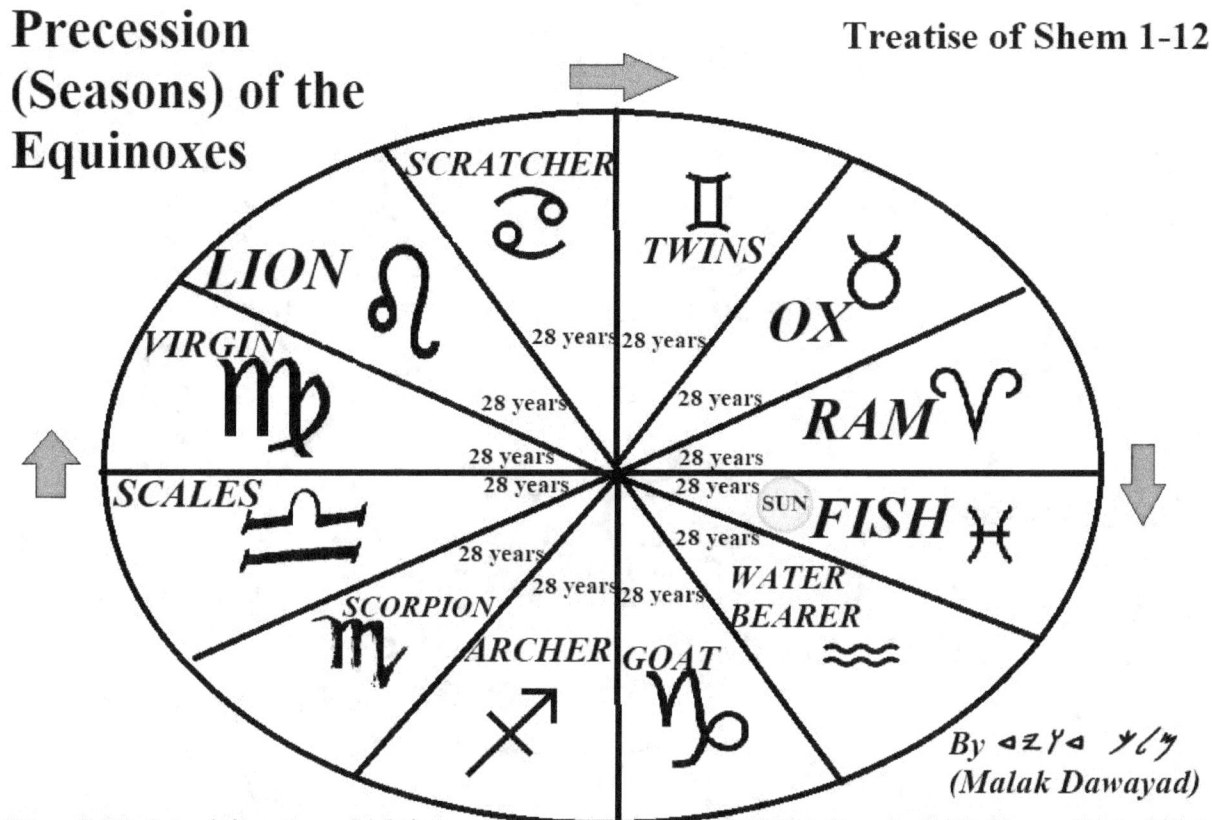

Precession (Seasons) of the Equinoxes

Treatise of Shem 1-12

By ᐊᙆᎩᐊ Ꭹᒪᘯ (Malak Dawayad)

2Enoch 15:3 And the gates which it (sun) enters, these are the great gates of the calculation of the hours of the year; for this reason the sun is a great creation, whose circuit (lasts) twenty-eight years, and begins again from the beginning.

2Enoch 21:7-8 And I saw the eighth (first) heaven, which is called in the Hebrew tongue Muzaloth, changer of the seasons ("Precessions of the Equinoxes"), of drought, and of wet, and of the twelve constellations (Zodiac) of the circle of the firmament, which are above the seventh heaven (7th world/Earth). And I saw the ninth (second) heaven, which is called in Hebrew Kuchavim, where are the heavenly homes of the twelve constellations of the circle of the firmament.

Treatise of Shem 1-12

Treatise of Shem 1:1 If the year begins in Aries...

Treatise of Shem 2:1 And if the year begins in Taurus...

Treatise of Shem 3:1 And if the year begins in Gemini: The moon will be beautiful and a north wind will blow...

Treatise of Shem 4:1 And if the year begins in Cancer: In the beginning of the year there will be a sufficiency of produce...
Treatise of Shem 5:1 And if the year begins in Leo...
Treatise of Shem 6:1 And if the year begins in Virgo...
Treatise of Shem 7:1 And if the year begins in Libra...
Treatise of Shem 8:1 And if the year begins in Scorpio: The north wind will blow in the beginning of the year...
Treatise of Shem 9:1 And if the year begins in Sagittarius...
Treatise of Shem 10:1 And if the year begins in Capricorn...
Treatise of Shem 11:1 And if the year begins in Pisces...
Treatise of Shem 12:1 When the year begins in Aquarius...

Ancient Hebrews observed the Calendar in tune with Enochian Astrology as the sun passes through each of the 12 signs. According to the apocryphal book in the 1st verse of each of the above chapters of the "Treatise of Shem," the Precession of the equinoxes (in the 3rd & 4th portals) isn't limited to the Astrological sign known as the Ram (Aries), which means the new year isn't locked into the Gregorian centered seasons and climate. The Enochian seasons are governed by the enochian portals (3rd & 4th) mentioned in 1Enoch 72 & 78 and the enochian astrological signs as they revolve around the first heaven known as "Muzaloth (Mazzaroth)" in 2Enoch 21:7-8

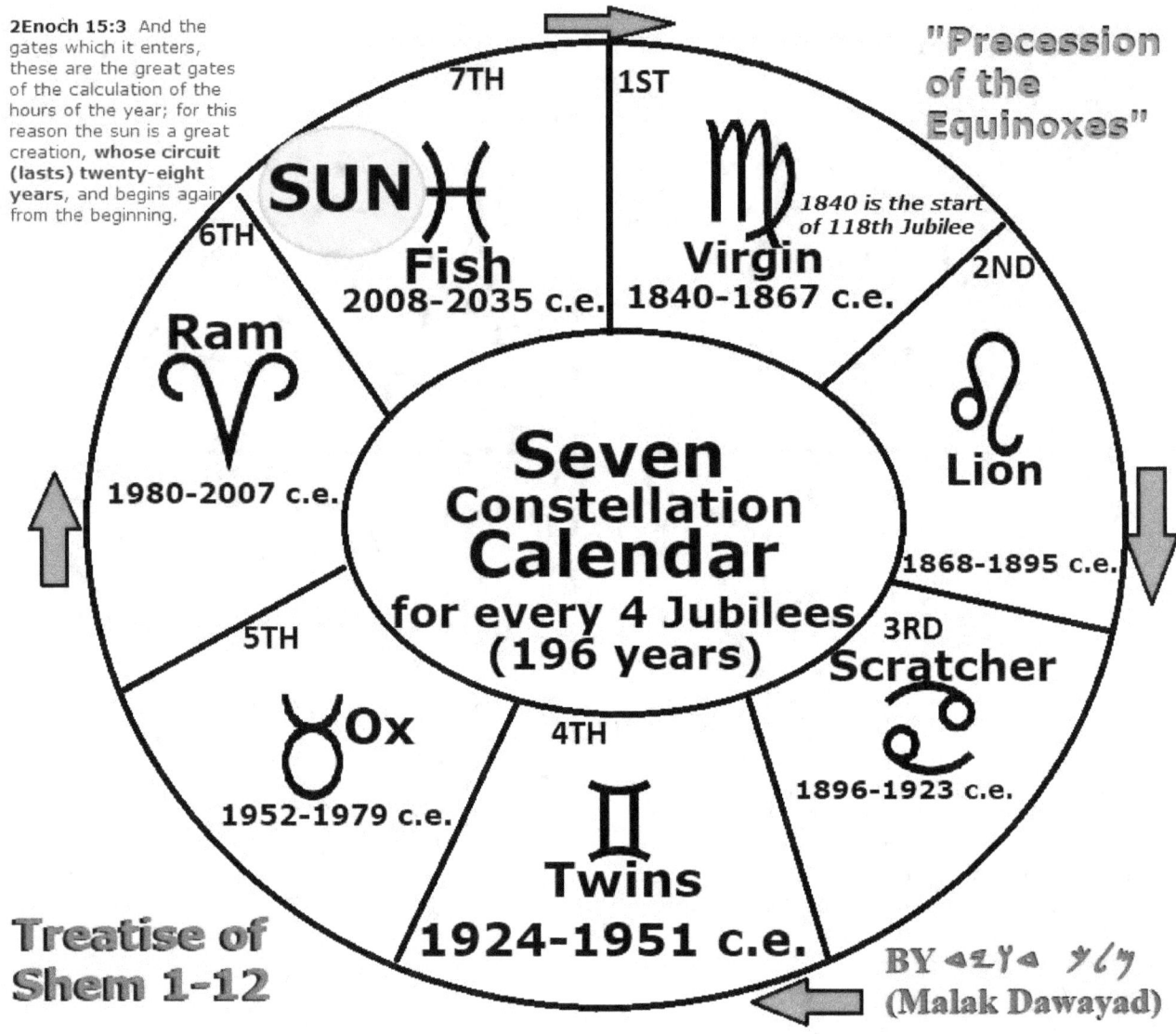

Being that each constellation lasts 28 years according to 2Enoch 15, a full cycle of 12 constellations would be 336 years. However, every 7 constellations equals the sum of exactly 4 complete jubilees. Once each of the 12 zodiac constellations is complete, the entire cycle simply repeats. This cycle revolves 3.5 times completing exactly 24 jubilees, thus repeating again, this cycle happens every 1,176 years. There's a larger cycle of 7 times completing exactly 48 jubilees, thus repeating again, this cycle happens every 2,352 years. I will incorporate this information partially in the jubilee calendar chapter for educational purposes.

jubilee- 49 years (Lev 25:8-22)

1 zodiac constellation- 28 years (2Enoch 15)

7 constellations- 196 years (4 jubilees)

full 12-zodiac cycle- 336 years

3.5 zodiac cycles- 1,176 yrs (24 jubilees)

7 zodiac cycles- 2,352 yrs (48 jubilees) (book of Jubilees 50:4)

FOR ENTERTAINMENT PURPOSES ONLY   BY ᚐᚎᚔᚐ ᚒᚓᚔ
(Malak Dawayad)

120th Jubilee

(1938 C.E.) 1 Palestinian revolts against Britain	**2** WWII	**Italy attacks 3** Palestine via Airstrike/ Death of Marcus Garvey	**4**	**5**	**6** Polish Army arrives in Palestine and later become Polish Jews	**(1944 C.E.) 7** Black Americans denied G.I. Bill by U.S. Gov after WWII ends..
(1945 C.E.) 8 WWII officially ends../ President H. Truman Requests 100,000 european refugees to enter Palestine from British gov	**9** Civil Rights Movement Began	**10** Khazar militia Palestine invasion	**11** Khazars settle in Palestine as "Jews"	**12**	**13**	**(1951 C.E.) 14** Chicago (Cicero), IL Riots
(1952 C.E.) 15 Age of the OX begins Egyptian President Nasser says "Jews left black and came back white, therefore cant be accepted"	**16**	**17**	**18** Vietnam War Alexandria Egypt Earthquake **Murder of Emmit Till**	**19** Lebanon Earthquake/ UK, France & Israel (Khazars) attack Egypt in October	**20**	**(1958 C.E.) 21** **Nasa Established**
(1959 C.E.) 22 Solar Eclipse/ Yellowstone Earthquake in Montana,U.S./ Antarctic Treaty signed by 12 nations	**23**	**24**	**25** Mt. Zion (Tel Arad, *Israel*) discovered/ uncoverd by Archaeologist Yohanan Ahroni	**26** **Murder of JFK**	**27**	**(1965 C.E.) 28** **Murder of Malcom X**
(1966 C.E.) 29 Ben Ammi's Vision of the Promised Land	**(Womans Lib) 30** American Hebrews settle in Liberia/ 6-day War Khazars Vs. Ottomans	**D.Gregory 4 President/ 31** Cointelpro Prevention of a Black Messiah by J.Edgar Hoover (Murder of RFK) **Murder of Dr. King**	**32** Egyptian Sinai Earthquake/ Murder of Fred Hampton/ American Hebrews settle near ancient Jerusalem (Arad, Israel)	**Shaleak Ben Yahuda 33** joins Hebrew Movement Khazarian Gov in Palestine changes "law of return," after American Hebrews arrive	**34** Prince Asiel Joins the American Hebrew Movement	**(1972 C.E.) 35** Ben Ammi leads remnant away from Arad to Dimona, Brother Job is mudered in the park in Dimona, Ben Ammi assumes leadership over entire American Hebrew Movement
(1973 C.E.) 36 Musician Abshalom Ben Shlomo joins Movement in Dimona/ Yom Kippur War Khazar Vs. Egypt & Syria	**37**	**38** Death of Elijah Muhammed/ **Vietnam War Ends**	**39**	**40** Farrakhan is crowned new leader of NOI, Ben Ammi was crowned in a public ceremony as King of Kings and Lord of Lords/ Louis Farrakhan & Dr. Khalid Muhammed Visit Arad (Ancient Mt.Olives) & Dimona	**41**	**(1979 C.E.) 42** Peace Treaty Signed between Israel & Egypt
(1980 C.E.) 43 Age of the RAM begins	**44**	**45**	**46** U.S. Gov recognizes Dr. M.L King Jr's Birthday as a U.S. Federal Holiday	**47** Rev. Jesse Jackson runs for U.S. presidency	**48** over 500 U.S. philadelphia police officers ambush and bomb a black miiddle-class residential neighborhood	**(1986 C.E.) 49** Khazarian gov raid in Dimona in the middle of the night seizing & deporting 46 members which lead to the famous "day of the show of strength"

THE ENOCH CALENDAR
1st month

The Sun is in the 4th Portal,
given 10 Enochian hrs (modern 13hrs & 20min) of light &
8 hrs (10hrs & 40min) of darkness by the 30th day

Leviticus 23:5-6

5.) In the fourteenth *day* of the first month at even *is* the LORD'S passover. 6.) And on the fifteenth day of the same month *is* the feast of unleavened bread unto the LORD: seven days ye must eat unleavened bread.

1st DAY	2nd DAY	3rd DAY	4th DAY	5th DAY	6th DAY	7th DAY
1st Jacobs Ladder Jubilees 27:19-27/Levi Day Jubilees 28	2nd	3rd	4th	5th	6th	7th 1st SABBATH
8th Lazarus Resurrection St. John 11-12	9th	10th Separate Pasacha Lamb, Exodus 12:3-6/ Death of Miriam, Jerahmeel 48:17	11th	12th	13th	14th 2nd SABBATH/**Passover** Death of Christ, Mat 27:50-64
15th Feast of Unleavened Bread	16th Feast of Unleavened bread	17th 3rd day of Feast/Christ Resurrection Luke 24:46	18th Feast of Unleavened bread	19th Feast of Unleavened bread	20th Feast of Unleavened bread	21st 3rd SABBATH/ Feast of Unleavened Bread
22nd Bring in sheaves to Milo, Lev 23:10-11	23rd	24th	25th	26th	27th	28th 4th SABBATH
29th	30th					

* All dates begin @twilight, **NOT** at 12am/midnight or sunrise
* The days of the month in bold numbers are Enoch calendar dates

THE ENOCH CALENDAR
2nd month

The Sun is in the 5th Portal,
given 11 Enochian hrs (modern 14hrs & 40min) of light &
7 hrs (9hrs & 20min) of darkness by the 30th day

Numbers 9:10-11

9.) Speak unto the children of Israel, saying, If any man of you or of your posterity shall be unclean by reason of a dead body, or *be* in a journey afar off, yet he shall keep the passover unto the LORD. 10.) The fourteenth day of the second month at even they shall keep it, *and* eat it with unleavened bread and bitter *herbs*.

1Maccabees 13:51-52

51.) And entered into it the three and twentieth day of the second month in the hundred seventy and first year, with thanksgiving, and branches of palm trees, and with harps, and cymbals, and with viols, and hymns, and songs: because there was destroyed a great enemy out of Israel. 52.) He ordained also that that day should be kept every year with gladness. Moreover the hill of the temple that was by the tower he made stronger than it was, and there he dwelt himself with his company.

1st DAY	2nd DAY	3rd DAY	4th DAY	5th DAY	6th DAY	7th DAY
		1st /31	2nd	3rd	4th	5th /35
			Death of Noah, Book of the Bee Chap.20			5th SABBATH
6th	7th	8th	9th	10th	11th	12th /42
						6th SABBATH
13th	14th /44	15th	16th	17th	18th	19th /49
	2nd Passover/ Unleavened	Unleaven/Rain of manna starts in wilderness, Exo 16:1-5	Unleavened Feast	Unleavened/Flood of Noah begins, Gen. 7:11, Jub. 5:23	Unleavened Feast	1st Legal Sabbath Exodus 16:23-30 Unleavened Feast / 7th SABBATH
20th	21st	22nd	23rd	24th	25th	26th /56
Unleavened Feast	Unleavened Feast	Bring in sheaves to Milo, Lev 23:10-11	Day of Simon's Victory over Tryphon King of Asia			8th SABBATH
27th	28th	29th	30th /60			

* All dates begin @twilight, **NOT** at 12am/midnight or sunrise
* The days of the month in bold numbers are Enoch calendar dates
* The back-slashed numbers on each Sabbath are the numbered days of the year

THE ENOCH CALENDAR
3rd month

The Sun is in the 6th Portal, given 12 Enochian hrs (modern 16hrs) of light & 6 hrs (8hrs) of darkness by the 31st day

Leviticus 23:15-17

15.) And ye shall count unto you from the morrow after the sabbath, from the day that ye brought the sheaf of the wave offering; seven sabbaths shall be complete:

16.) Even unto the morrow after the seventh sabbath shall ye number fifty days; and ye shall offer a new meat offering unto the LORD. 17.) Ye shall bring out of your habitations two wave loaves of two tenth deals: they shall be of fine flour; they shall be baken with leaven; *they are* the firstfruits unto the LORD.

1st DAY	2nd DAY	3rd DAY	4th DAY	5th DAY	6th DAY	7th DAY
				1st /61	2nd	3rd /63
						Moses & Israelites arrive @Mt.Sinai, Jerahmeel 48:16 9th SABBATH
4th	5th	6th	7th	8th	9th	10th /70
		Enoch Birth ascends, 2nd Enoch 68/ Law Given to Moses, Jerahmeel 48:16				Legends of the Jews Book 4, Chapter 4: Death of David 10th SABBATH
11th /71	12th	13th	14th	15th /75	16th	17th /77
Feast of First Fruits/Weeks		Angels visit Abram before destroying Sodom Jasher 18:1-9		Patriarch Judah Day, Jubilees 28	Death of Adam 2Adam&Eve 9:1-6	Summer Solstice 2nd Enoch 48:1-2 11th SABBATH
18th	19th	20th	21st	22nd	23rd	24th /84
						12th SABBATH
25th	26th	27th	28th	29th	30th	31st /91
						13th SABBATH/ Enoch 82:4-14 Intercalary Day #1

* All dates begin @twilight, NOT at 12am/midnight or sunrise
* The days of the month in bold numbers are Enoch calendar dates
* The back-slashed numbers on each Sabbath are the numbered days of the year

THE ENOCH CALENDAR
4th month

The Sun is in the 6th Portal,
given 11 Enochian hrs (modern 14hrs & 40min) of light &
7 hrs (9hrs & 20min) of darkness by the 30th day

<u>2Kings 25:3-7</u> And on the ninth *day* of the *fourth* month the famine prevailed in the city, and there was no bread for the people of the land. And the city was broken up, and all the men of war *fled* by night by the way of the gate between two walls, which *is* by the king's garden: (now the Chaldees *were* against the city round about:) and *the king* went the way toward the plain. And the army of the Chaldees pursued after the king, and overtook him in the plains of Jericho: and all his army were scattered from him. So they took the king, and brought him up to the king of Babylon to Riblah; and they gave judgment upon him. And they slew the sons of Zedekiah before his eyes, and put out the eyes of Zedekiah, and bound him with fetters of brass, and carried him to Babylon.

* All dates begin @twilight, NOT at 12am/midnight or sunrise
* The days of the month in bold numbers are Enoch calendar dates
* The back-slashed numbers on each Sabbath are the numbered days of the year

1st DAY	2nd DAY	3rd DAY	4th DAY	5th DAY	6th DAY	7th DAY
1st /92	2nd	3rd	4th	5th	6th	7th /98
New Month/ Joseph Day Jubilees 28				3-Day War against Ptolemy, 3Macc 5-6	War against Ptolemy	Angels Victory against Ptolemy 3Macc 6:16-28/ 14th SABBATH
8th	9th	10th /101	11th	12th	13th	14TH /105
	Nebuchadenezzar captures King Zedekiah 2ki 25:3-7	Fast of the 4th Month, Zec 8:19		Death of Jared 2Adam&Eve 21:12-14		15th SABBATH
15th	16th	17th	18th	19th	20th	21st /112
						16th SABBATH
22nd	23rd	24th	25th	26th	27th	28th /119
						Patriarch's Day, Testament of Isaac/Jacob 1:1-3, 8:1-2 17th SABBATH
29th	30th /121					

THE ENOCH CALENDAR
5th month

The Sun is in the 5th Portal, given 10 Enochian hrs (modern 13hrs & 20min) of light & 8 hrs (10hrs & 40min) of darkness by the 30th day

Numbers 33:38

And Aaron the priest went up into mount Hor at the commandment of the LORD, and died there, in the fortieth year after the children of Israel were come out of the land of Egypt, in the first *day* of the fifth month.

1st DAY	2nd DAY	3rd DAY	4th DAY	5th DAY	6th DAY	7th DAY
		1st /122	2nd	3rd	4th /125	5th /126
		Death of 1st Priest Aaron			Issachar Day Jubilees 28	18th SABBATH
6th	7th	8th	9th	10th	11th	12th /133
	Nebuchadenezzar burns temple and Jerusalem, 2kings 25:8-21	Fast of the 5th Month, Zec 8:19				19th SABBATH
13th	14th	15th	16th	17th	18th	19th /140
						20th SABBATH
20th	21st	22nd	23rd	24th	25th	26th /147
						21st SABBATH
27th	28th	29th	30th /151			

* All dates begin @twilight, NOT at 12am/midnight or sunrise

* The days of the month in bold numbers are Enoch calendar dates

* The back-slashed numbers on each Sabbath are the numbered days of the year

THE ENOCH CALENDAR
6th Month

The Sun is in the 4th Portal,
given 9 Enochian hrs (modern 12hrs) of light &
9 hrs (12hrs) of darkness by the 31st day
(Equinox/Equal day & night)

Jerahmeel 61:1

The banishment brought about by Titus, Vespasianus, and Hadrian, occurred on the eve of the ninth of Ab, on the outgoing of the Sabbath and the Sabbatical year. The Levites were then occupied with their ministrations, and with their harps in their hands, were singing their hymns. But scripture saith, 'He hath brought upon them their own iniquity, and shall cut them off in their own evil.' The words 'He shall cut them off' were not yet fully uttered ere their enemies came upon them, slaughtered many of them, and sent the rest into exile...

* All dates begin @twilight, NOT at 12am/midnight or sunrise
* The days of the month in bold numbers are Enoch calendar dates
* The back-slashed numbers on each Sabbath are the numbered days of the year

1st DAY	2nd DAY	3rd DAY	4th DAY	5th DAY	6th DAY	7th DAY
				1st/152 New Month 1Sa 20:5,25-34	2nd	3rd/154 David eats hallowed bread from temple, 1Sam 21:1-9 Matt 12:2-8, Lev 24:5-9 22nd SABBATH
4th	5th Ezekiel's vision of Angel pulling him by one of his locks, Ezek 8:1-7	6th	7th	8th	9th/160 2nd Temple destroyed Jerahmeel 61:1 Dan's Day Jubilees 28	10th/161 23rd SABBATH
11th	12th	13th	14th	15th	16th	17th/168 New Oil, Remembrance of Enoch Ch. 13 24th SABBATH
18th High Priest Simon honored in Zion, 1Mac 14:25-27	19th	20th	21st	22nd	23rd	24th/175 25th SABBATH
25th	26th	27th	28th	29th	30th	31st/182 1st Equinox 26th SABBATH/ Enoch 82:4-14 Intercalary Day #2

THE ENOCH CALENDAR
7th Month

The Sun is in the 3rd Portal, given 8 Enochian hrs (modern 10hrs & 40min) of light & 10 hrs (13hrs & 20min) of darkness by the 30th day

Leviticus 23:34
Speak unto the children of Israel, saying, The fifteenth day of this seventh month *shall be* the feast of tabernacles *for* seven days unto the LORD.

1st DAY	2nd DAY	3rd DAY	4th DAY	5th DAY	6th DAY	7th DAY
1st/183	2nd	3rd	4th	5th	6th	7th/189
Memorial of Trumpets Lev 23:24-25				Patriarch Naphtali Day Jubilees 28		Zebulon & Dinah Day, Jubilees 28/ 27th SABBATH
8th	9th	10th/192	11th	12th	13th	14th/196
	Fast begins @ even	Atonement/ Joseph sold by his brothers, Jubilees 34:10-19				28th SABBATH
15th	16th	17th	18th	19th	20th	21st/203
Feast of Tabernacles	Feast of Tabernacles	Feast of Tabernacles	Feast of Tabernacles	Feast of Tabernacles	Feast of Tabernacles	29th SABBATH/ Feast of Tabernacles
22nd	23rd	24th	25th	26th	27th	28th/210
Jacob wrestles with Angel Jubilees 32:16-29/ Tabernacles Feast						30th SABBATH
29th	30th/212					

* All dates begin @twilight, NOT at 12am/midnight or sunrise
* The days of the month in bold numbers are Enoch calendar dates
* The back-slashed numbers on each Sabbath are the numbered days of the year

THE ENOCH CALENDAR
8th Month

The Sun is in the 2nd Portal,
given 7 Enochian hrs (modern 9hrs & 20min) of light &
11 hrs (14hrs & 40min) of darkness by the 30th day

Jubilees 32:33-34

33. And Rachel bare a son in the night, and called his name "Son of my sorrow"; for she suffered in giving him birth: but his father called his name Benjamin, on the eleventh of the eighth month in the first of the sixth week of this jubilee.

34. And Rachel died there and she was buried in the land of Ephrath, the same is Bethlehem, and Jacob built a pillar on the grave of Rachel, on the road above her grave.

1st DAY	2nd DAY	3rd DAY	4th DAY	5th DAY	6th DAY	7th DAY
		1st /213	2nd	3rd	4th	5th /217
						31st SABBATH
6th	7th	8th	9th	10th	11th /223	12th /224
					Rachel's Death/ Benjamin Day	Patriarch Gad Day, Jubilees 28/32nd SABBATH
13th	14th	15th /227	16th	17th	18th	19th /231
		Jeroboam's Pagan Feast to the Golden Calf 1kings 12:32-33			Death of Ezra, Greek Apocalypse of Ezra 7:14-15	33rd SABBATH
20th	21st	22nd	23rd	24th	25th	26th /238
						34th SABBATH
27th	28th	29th	30th /242			

* All dates begin @twilight, NOT at 12am/midnight or sunrise
* The days of the month in bold numbers are Enoch calendar dates
* The back-slashed numbers on each Sabbath are the numbered days of the year

THE ENOCH CALENDAR
9th Month

The Sun is in the 1st Portal,
given 6 Enochian hrs (modern 8hrs) of light &
12 hrs (16hrs) of darkness by the 31st day

1Maccabees 4:59

59.) Moreover Judas and his brethren with the whole congregation of Israel ordained, that the days of the dedication of the altar should be kept in their season from year to year by the space of eight days, from the five and twentieth day of the month Casleu (9th month), with mirth and gladness.

1st DAY	2nd DAY	3rd DAY	4th DAY	5th DAY	6th DAY	7th DAY
				1st/243	2nd	3rd/245
						35th SABBATH
4th	5th	6th	7th	8th	9th	10th/252
						36th SABBATH
11th	12th	13th	14th/256	15th	16th	17th/259
Jacob & Esau Reconcile, Jubilees 29:13			Patriarch Reuben Day Jubilees 28			Winter Solstice 2nd Enoch 48:1-2 37th SABBATH
18th	19th	20th	21st	22nd	23rd	24th/266
						38th SABBATH
25th/267	26th	27th	28th	29th	30th	31st/273
Feast of Dedication	Feast of Dedication	Feast of Dedication	Feast of Dedication	Feast of Dedication	Feast of Dedication	39th SABBATH/ Feast of Dedication/Intercalary Day #3 Enoch 82:4-14

* All dates begin @twilight, NOT at 12am/midnight or sunrise
* The days of the month in bold numbers are Enoch calendar dates
* The back-slashed numbers on each Sabbath are the numbered days of the year

THE ENOCH CALENDAR

10th Month

The Sun is in the 1st Portal,
given 7 Enochian hrs (modern 9hrs & 20min) of light &
11 hrs (14hrs & 40min) of darkness by the 30th day

2Kings 25:1

And it came to pass in the ninth year of his reign, in the tenth month, in the tenth *day* of the month, *that* Nebuchadnezzar king of Babylon came, he, and all his host, against Jerusalem, and pitched against it; and they built forts against it round about.

* All dates begin @twilight, NOT at 12am/midnight or sunrise
* The days of the month in bold numbers are Enoch calendar dates
* The back-slashed numbers on each Sabbath are the numbered days of the year

1st DAY	2nd DAY	3rd DAY	4th DAY	5th DAY	6th DAY	7th DAY
1st/274	2nd	3rd	4th	5th	6th	7th/280
New Month, Dedication Feast, 1Mac 4:59						40th SABBATH
8th	9th	10th	11th	12th	13th	14th/287
		Siege of Jerusalem /10th mth fast, Zec 8:19				41st SABBATH
15th	16th	17th	18th	19th	20th	21st/294
						Simeon Day Jubilees 28/ 42nd SABBATH
22nd	23rd	24th	25th	26th	27th	28th/301
						43rd SABBATH
29th	30th/303					

THE ENOCH CALENDAR
11th Month

The Sun is in the 2nd Portal,
given 8 Enochian hrs (modern 10hrs & 40min) of light &
10 hrs (13hrs & 20min) of darkness by the 30th day

Deuteronomy 1:3,6-7

Deu 1:3 And it came to pass in the fortieth year, in the eleventh month, on the first *day* of the month, *that* Moses spake unto the children of Israel, according unto all that the LORD had given him in commandment unto them;

Deu 1:6 The LORD our God spake unto us in Horeb, saying, Ye have dwelt long enough in this mount:

Deu 1:7 Turn you, and take your journey, and go to the mount of the Amorites, and unto all *the places* nigh thereunto, in the plain, in the hills, and in the vale, and in the south, and by the sea side, to the land of the Canaanites, and unto Lebanon, unto the great river, the river Euphrates.

* All dates begin @twilight, NOT at 12am/midnight or sunrise
* The days of the month in bold numbers are Enoch calendar dates
* The back-slashed numbers on each Sabbath are the numbered days of the year

1st DAY	2nd DAY	3rd DAY	4th DAY	5th DAY	6th DAY	7th DAY
		1st /304	2nd /305	3rd	4th	5th /308
		The Exodus journeys from Horeb	Asher Day, Jubilees 28			44th SABBATH
6th	7th	8th	9th	10th	11th	12th /315
						45th SABBATH
13th	14th	15th	16th	17th	18th	19th /322
						46th SABBATH
20th	21st	22nd	23rd	24th	25th	26th /329
				Horseman & 4 horns Vision, Zec 1:7-21		47th SABBATH
27th	28th	29th	30th /333			

THE ENOCH CALENDAR
12th Month

The Sun is in the 3rd Portal,
given 9 Enochian hrs (modern 12hrs) of light &
9 hrs (12hrs) of darkness by the 31st day
(Equinox/Equal day & night)

* All dates begin @twilight, NOT at 12am/midnight or sunrise
* The days of the month in bold numbers are Enoch calendar dates
* The back-slashed numbers on each Sabbath are the numbered days of the year

1Maccabees 7:47-49

Afterwards they took the spoils, and the prey, and smote off Nicanors head, and his right hand, which he stretched out so proudly, and brought them away, and hanged them up toward Jerusalem. For this cause the people rejoiced greatly, and they kept that day a day of great gladness. Moreover they ordained to keep yearly this day, being the thirteenth of Adar (12th month).

Esther 9:21-22

To stablish this among them, that they should keep the fourteenth day of the month Adar (12th month), and the fifteenth day of the same, yearly, As the days wherein the Jews rested from their enemies, and the month which was turned unto them from sorrow to joy, and from mourning into a good day: that they should make them days of feasting and joy, and of sending portions one to another, and gifts to the poor.

1st DAY	2nd DAY	3rd DAY	4th DAY	5th DAY	6th DAY	7th DAY
				1st /334	2nd	3rd /336 48th SABBATH
4th	5th	6th	7th /340 Birth & Death of Moses, Jerahmeel 48:17	8th	9th	10th /343 49th SABBATH
11th	12th	13th /346 Day of Nicanor	14th /347 Feast of Purim	15th /348 Feast of Purim	16th	17th /350 50th SABBATH
18th	19th	20th	21st	22nd	23rd	24th /357 51st SABBATH
25th Angel appears to Mary, Book of the Bee Ch.35:1	26th	27th	28th	29th	30th	31st /364 52nd SABBATH/ Equinox #2 **Intercalary Day #4** Enoch 82:4-14

LITERATURE THAT SUPPORTS THE ENOCH CALENDAR

BOOKS OF ENOCH 1-2
REMEMBRANCE OF ENOCH
TORAH (OT)
PROPHETS (OT)
APOCRYPHA (OT+)
NEW TESTAMENT (NT)
JUBILEES
JASHER
LEGENDS OF THE JEWS
TREATIES OF SHEM
CHRONICLES OF JERAHMEEL
BOOK OF THE BEE
1st & 2nd ADAM & EVE

THE UNIVERSAL ENOCH CALENDAR FROM THE REAL JERUSALEM IN TEL ARAD, ISRAEL BY ◁ᴢＹ◁ Ｙ⌊ᴍ (Malak Dawayad)

Psalms 90:12 So teach *us* to number our days, that we may apply *our* hearts unto wisdom.

Please Donate to Continue the Research:

$malakdavid

supaproducerbeats@gmail.com

Contact

Dawayad Baba Zion

malakdavid21

Malak Dawayad

supaproducerbeats@gmail.com

www.ingramcontent.com/pod-product-compliance
Lightning Source LLC
Chambersburg PA
CBHW062316220526
45479CB00004B/1198